Understanding Breast Cancer

Cell Biology and Therapy—A Visual Approach

Colloquium Series on the Cell Biology of Medicine

Editor
Joel D. Pardee, CEO, Neural Essence LLC; formerly Associate Professor and Associate Dean of Graduate Research, Weill Cornell Medical College

In order to learn we must be able to remember, and in the world of science and medicine we remember what we envision, not what we hear. It is with this essential precept in mind that we offer the Cell Biology of Medicine series. Each book is written by faculty accomplished in teaching the scientific basis of disease to both graduate and medical students. In this modern age it has become abundantly clear that everyone is vastly interested in how our bodies work and what has gone wrong in disease. It is likewise evident that the only way to understand medicine is to engrave in our mind's eye a clear vision of the biological processes that give us the gift of life. In these lectures, we are dedicated to holding up for the viewer an insight into the biology behind the body. Each lecture demonstrates cell, tissue and organ function in health and disease. And it does so in a visually striking style. Left to its own devices, the mind will quite naturally remember the pictures. Enjoy the show.

Published

Cell Origin, Structure and Function: How Cells Make a Living—A Visual Approach
Joel D. Pardee
2010

The Actin Cytoskeleton in Cell Motility, Cancer, and Infection
Joel D. Pardee
2010

Skeletal Muscle & Muscular Dystrophy: A Visual Approach
Donald A. Fischman
2009

How the Heart Develops: A Visual Approach
Donald A. Fischman
2009

Forthcoming Titles:

Enhancement Therapy for Mental Disorders
Joel D. Pardee
2011

The Body Plan: How Structure Creates Function
Joel D. Pardee
2011

Bones: Growth, Strength, and Osteoporosis
Michelle Fuortes
2011

Cancer Genetics
A.M.C. Brown
2011

Cancer Invasion and Metastasis
A.M.C. Brown
2011

Cartilage: Keeping Joints Functioning
Michelle Fuortes
2011

The Cell Cycle and Cell Division
Joel D. Pardee and A.M.C. Brown
2011

Cell Metabolic Enhancement Therapy for Mental Disorders
Joel D. Pardee
2011

Cell Motility in Cancer and Infection
Joel D. Pardee
2011

Cell Transformation and Proliferation in Cancer
A.M.C. Brown
2011

Cholesterol and Complex Lipids in Medicine: Membrane Building Blocks
Suresh Tate
2011

Creating Proteins from Genes
Phil Leopold
2011

Development of the Cardiovascular System
D.A. Fischman
2011

Digestion and the Gut: Exclusion and Absorption
Joel D. Pardee
2011

The Extracellular Matrix: Biological Glue
Joel D. Pardee and A.M.C. Brown
2011

Fertilization, Cleavage, and Implantation
D.A. Fischman
2011

Gastrulation, Somite Formation, and the Vertebrate Body Plan
D.A. Fischman
2011

Generating Energy by Oxidative Pathways: Mitochondrial Power
Suresh Tate
2011

Heart Structure and Function: Why Hearts Fail
D.A. Fischman
2011

Hematopoiesis and Leukemia
Michelle Fuortes
2011

How Amino Acids Create Hemoglobin, Neurotransmitters, DNA, and RNA
Suresh Tate
2011

The Human Genome and Personalized Medicine
Phil Leopold
2011

Mechanisms of Cell & Tissue Aging: Why We Get Old
Joel D. Pardee
2011

Mendelian Genetics in Medicine
Phil Leopold
2011

Metabolism of Carbohydrates: Glucose Homeostasis in Fasting and Diabetes
Suresh Tate
2011

Metabolism of Fats: Energy in Lipid Form
Suresh Tate
2011

Metabolism of Protein: Fates of Amino Acids
Suresh Tate
2011

Neurulation, Formation of the Central and Peripheral Nervous Systems
D.A. Fischman
2011

Non-Mendelian Genetics in Medicine
Phil Leopold
2011

Skeletal Muscle, Muscular Dystrophies, and Myastheneis Gravis
D.A. Fischman
2011

Skin: Keeping the Outside World Out
Joel D. Pardee
2011

Stem Cell Biology
A.M.C. Brown
2011

Therapies for Genetic Diseases
Phil Leopold
2011

Tissue Regeneration: Renewing the Body
Joel D. Pardee
2011

Understanding Breast Cancer: Cell Biology and Therapy—A Visual Approach
Joel D. Pardee
www.morganclaypool.com

ISBN: 9781615040759 paperback

ISBN: 9781615040766 ebook

DOI: 10.4199/C00021ED1V01Y201012CBM004

A Publication in the Morgan & Claypool Life Sciences series

COLLOQUIUM SERIES ON THE CELL BIOLOGY OF MEDICINE

Book #5

Series Editor: Joel D. Pardee, CEO, Neural Essence LLC and formerly Associate Dean and Associate Professor, Weill Cornell Medical College

Series ISSN

ISSN 2153-0513 print

ISSN 2153-0521 electronic

Understanding Breast Cancer

Cell Biology and Therapy—A Visual Approach

Joel D. Pardee
CEO, Neural Essence LLC and formely Associate Dean and
Associate Professor, Weill Cornell Medical College

COLLOQUIUM SERIES ON THE CELL BIOLOGY OF MEDICINE #5

MORGAN & CLAYPOOL LIFE SCIENCES

ABSTRACT

The mysterious disease of cancer, including breast cancer, has plagued mankind since the dawn of recorded history. Regarding the elusive cause of the disease, the "Father of Medicine," Hippocrates of Athens (460–377 BC), wrote that, "For instability is characteristic of the humours and so they may be easily altered by nature and by chance." The enigma has persisted until today. In 1971, then President Richard Nixon signed the National Cancer Act and declared a "War on Cancer." He believed the counsel of scientists and physicians that if sufficient resources were committed to the fight, cancer could be virtually eliminated within 5 years. The prophesy failed. Although mortality from a few cancers, most notably leukemias, has been significantly reduced, carcinomas, cancers of the epithelium, which account for 80% of cancer deaths, remain unchanged. While tremendous advances have taken place in our understanding of the molecular and cellular mechanisms operant in cancer, it has proven exceedingly difficult to prevent the occurrence or to halt the progress of the disease. The very best therapy remains early detection while the primary tumor is small and localized to a single site, followed by removal of the offending growth by surgery and/or radiation. The great challenge of finding a cure confronts us yet, and it is effective intervention at the molecular level that offers our best hope. We still must find the "magic bullet."

KEYWORDS
breast cancer, carcinoma, therapies, origin, metastasis, estrogen receptor

Contents

The Problem of Cancer

We dance round in a ring and suppose.
But the secret sits in the middle and knows.
R. Frost

The mysterious disease of cancer has plagued mankind since the dawn of recorded history. Regarding the elusive cause of the disease, the "Father of Medicine," Hippocrates of Athens (460–377 BC), wrote that, "For instability is characteristic of the humours and so they may be easily altered by nature and by chance." The enigma has persisted until today. In 1971, then President Richard Nixon signed the National Cancer Act and declared a "War on Cancer." He believed the counsel of scientists and physicians that if sufficient resources were committed to the fight, cancer could be virtually eliminated within 5 years. The prophesy failed. Although mortality from a few cancers, most notably leukemias, has been significantly reduced, carcinomas, cancers of the epithelium, which account for 80% of cancer deaths, remain unchanged (Figure 1).

While tremendous advances have taken place in our understanding of the molecular and cellular mechanisms operant in cancer, it has proven exceedingly difficult to prevent the occurrence or to halt the progress of the disease. The very best therapy remains early detection while the primary tumor is small and localized to a single site, followed by removal of the offending growth by surgery and/or radiation. The great challenge of finding a cure confronts us yet, and it is effective intervention at the molecular level that offers our best hope. We still must find the "magic bullet."

TOTAL ESTIMATED CANCER DEATHS IN THE UNITED STATES		
SITE	1971	2002
Total cancer deaths in men		
Lung	53,000	89,000
Colorectal	22,000	28,000
Prostate	17,000	30,000
Pancreas	10,000	15,000
Total cancer deaths in women		
Lung	11,000	66,000
Colorectal	24,000	29,000
Breast	31,000	40,000
Pancreas	8,000	15,000
Ovary	10,000	14,000

Numbers are from American Cancer Society statistics. They have not been adjusted for age or population; the latter has increased by ~40% during the past 30 years. These data are not intended to imply that there have not been significant advances in the management of carcinoma at all of the above sites.

FIGURE 1: Cancer Mortalities.

How Carcinomas Become Lethal

Before we delve into the intricacies of breast cancer, we may well want to first ask the compelling question, "Why does cancer kill us?" The answer is best learned by visualizing the progression of cancer from onset of a single-cell transformation to terminal disseminated disease (Figure 2).

STAGE I - TRANSFORMATION

Cancer begins with the conversion of a single cell from a "normal" phenotype into a perpetually dividing "transformed" phenotype. A carcinoma is a cancer that originates in an epithelial tissue, the cells that line the intestinal tract, the heart and vasculature, lung, all ducts of the body (including breast lactation ducts), and the outermost layer of the skin (epidermis). The single transformed cell proliferates by unchecked cell division to form a *primary tumor* that is confined to the epithelial tissue of origin. Medical diagnosis refers to this state of development as Stage I (proliferation).

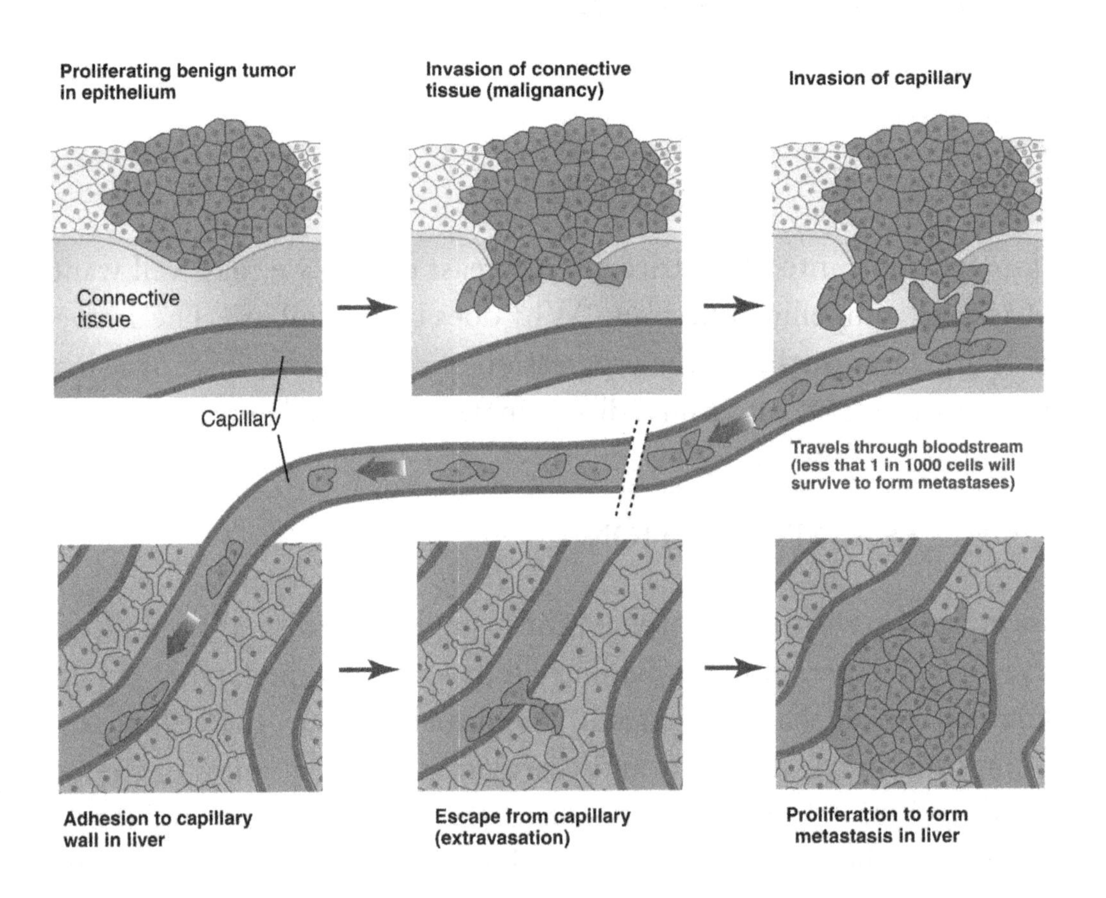

FIGURE 2: Carcinoma Tumor Progression. Based on the figure 23–15 from Molecular Biology of the Cell, 4th ed./Garland.

STAGE II - INVASION

As the tumor grows, a few tumor cells can be further transformed into a motile, invasive phenotype that is capable of migrating out of the tumor mass into the adjacent tissues. This is the event of tumor invasion. It is called Stage II, *Invasive Carcinoma*. In the example shown in Figure 2, a primary tumor has formed in the epidermis of the skin, and tumor cells have metastasized into the underlying connective tissue containing fibroblasts, immune system cells, nerves, and blood vessels embedded in an extracellular matrix of structural proteins. In the case of breast cancer, a primary tumor arises in the epithelial layer lining the lactation ducts and metastasizes into the underlying connective tissue surrounding the duct. It is worth pointing out that primary tumors that have not acquired metastatic potential are considered *benign* and are not lethal. It is through the conversion to the invasive phenotype that carcinoma cancers become *malignant* because they are capable of disseminating throughout the body.

STAGE III - METASTASIS

Once malignant cells invade the connective tissue, they can enter the bloodstream by migrating through the vessel wall, a process termed *diapedisis* (leaping through). Once in the circulation, cancer cells are swept through the vasculature (Stage III, *Metastasis*). At this stage, the immune system cells residing in blood and lymph nodes have direct access to the malignant cancer cells and can mount an immune response against them. In healthy individuals, the immune response can be quite effective, eliminating 95%–100% of circulating metastatic cells.

STAGE IV - DISSEMINATED DISEASE

In the final stage of cancer progression, malignant cells adhere to the vessel wall at preferred body sites that are determined by the type of carcinoma. For instance, breast cancer preferentially spreads to the brain and bone, while colon cancer goes to the liver. At the preferential site, cells retraverse the vessel wall by *extravasation* to enter the connective tissue of the target organ. Proliferation begins anew in the target organ to create *micrometastases*, new tumors. This is Stage IV, *Disseminated Disease*. It is the profligate expansion of tumor masses throughout the body that is lethal. The massive numbers of tumor cells that are generated capture the metabolic energy needed by normal tissue cells to survive, so that tissue and organ failure eventually result. In the final stages of the disease, metabolic starvation (*cachexia*) of normal tissues occurs to such an extent that the body becomes wasted. Muscle and fat disappear, organs become functionally impaired, and victims appear to be starving to death, which is true. The mortal blow is metabolic starvation.

Understanding the molecular details of cancer progression is crucial to therapy development. Each stage presents opportunities for intervention. For example, Stage IV (disseminated disease) indicates the use of antimitotic therapies that prevent tumor growth, which is, in fact, the basis for currently employed chemotherapeutics. Stage II (invasion) suggests development of cell motility inhibitors, and Stage III (metastasis) speaks to the development of immune enhancers.

Breast Cancer

EARLY WARNING

Breast cancer is the most frequently diagnosed cancer in American women. It is diagnosed in approximately 250,000 women per year in the United States and caused the death of 40,000 women in 2002 (see Fig. 1). Fifty percent of those diagnosed will have no recurrence if properly treated. Fifty percent will have a recurrence of the disease even after treatment, and approximately 30% of those in which recurrence has occurred will die of the disease. The increase in annual mortality from 31,000 in 1971 to 40,000 in 2002 reflects an increased population over the 30-year interval but clearly points out a failure to significantly reduce breast cancer deaths over the same period. Since patients with Stage I or II disease in which tumors are not associated with metastases to lymph nodes or distant sites show from 66% (Stage II) to 85% (Stage I) survival rates, it is clear that the earliest possible detection of ductal neoplasias remains the key to successful therapeutic intervention. Consequently, molecular markers capable of detecting ductal neoplasias before invasive tumor formation remain a goal for significantly improving survival rates in the general population.

TUMOR GROWTH

How long does it take for breast cancer to develop from initial transformation of a single cell into an invasive metastasis and disseminated disease? Although cases vary, the somewhat surprising answer is from 7 to 17 years. Breast cancer is a slowly developing disease (Figure 3), which agrees with the increasing awareness that carcinomas are a disease of aging. The year-long period for a single cell to reach detectable tumor size (roughly 0.5 to 1 cm diameter) seems odd, given the 30 cell doublings needed to reach that size and a doubling time of approximately 24 h for epithelial cells in culture. Thirty doublings should take a mere 30 days by that criterion. Tumors apparently grow much more slowly in the breast than in culture. Several reasons for this phenomenon have been posited. Perhaps most important is the genetic complexity of multistep carcinogenesis. As tumors develop, a series of five to seven separate genetic events have been identified that must occur in succession for cells to progress from transformation to metastasis. This takes time because each event is a spontaneous occurrence, not triggered by the previous genetic change. Intensive research effort over the past 20 years has succeeded in revealing a large number of cellular features characteristic of epithelial neoplasias, including activation of oncogenes, inhibition of tumor suppressor gene products, expression of steroid receptors, growth factors, and cell surface glycoproteins.

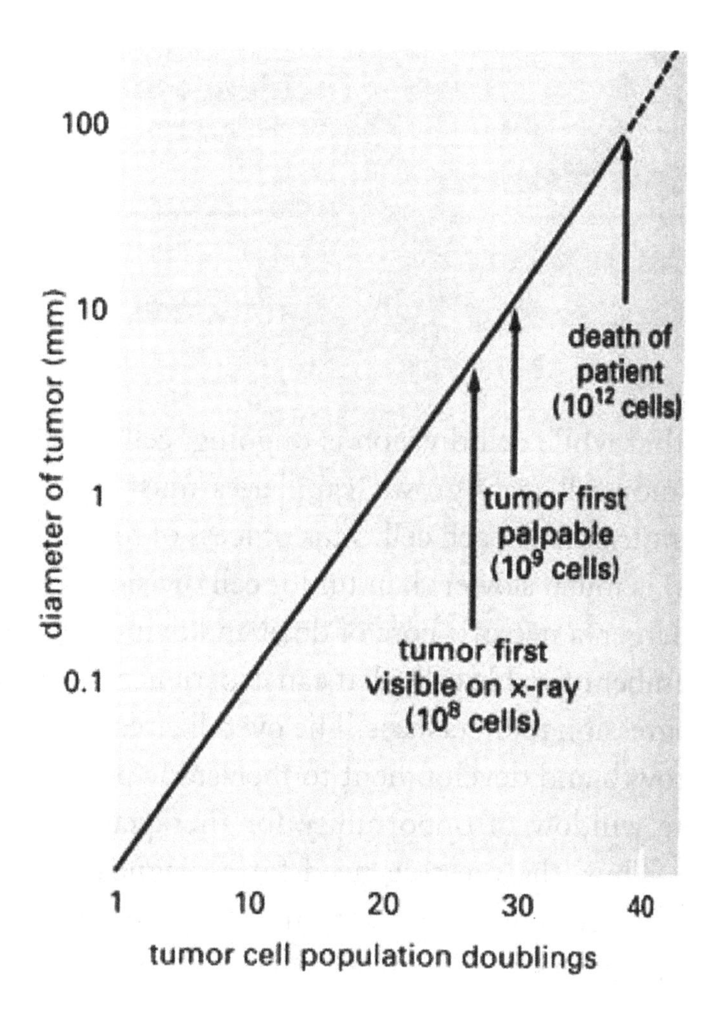

FIGURE 3: Rate of Tumor Growth. The diameter of a typical tumor of the breast is shown as a result of the number of tumor cell population doublings. Years may elapse before the tumor reaches a size that is detectable. The initial transformed cell must divide approximately 30 times before it can be detected by current methods. With only 10 more doublings, the tumor load is sufficient to be lethal. The elapsed time between detection and lethality is much shorter than the time between initial transformation and detection. Reproduced from *Molecular Biology of the Cell*, 3rd ed., Garland Science. Permission pending.

It is clear that while cell division is ongoing, cell death is also occurring. As the tumor cell mass grows, capillaries must invade the forming tumor to bring nutrients to each cell. This process of forming new capillaries (*angiogenesis*) is much slower than tumor cell division. Consequently, as the tumor gets larger, a necrotic core of dead and dying cells develops in it, reducing the number of viable cells that can experience the genetic changes required for progression to metastasis. The overall effect is a dramatic slowing of tumor growth and development to metastatic potential. This opens an inviting time window of opportunity for therapeutically intercepting tumor progress. Thus, the pressing need for a diagnostic that can detect individual transformed cells and microtumors at the onset of breast cancer. Such a fail-safe diagnosis is not yet available.

MAMMARY GLAND ANATOMY

Each mammary gland contains a series of branched lactation ducts that drain into the nipple (Figure 4). Budding from each lactation duct are a number of alveoli in which mother's milk and colostrum (first milk) is produced. Milk secreted by secretory epithelial cells lining the alveolus passes through the lactation duct to the nipple. Each lactation duct is lined with a layer of nonsecretory epithelial cells. It is primarily ductal epithelial cells that are transformed during the genesis of mammary carcinomas.

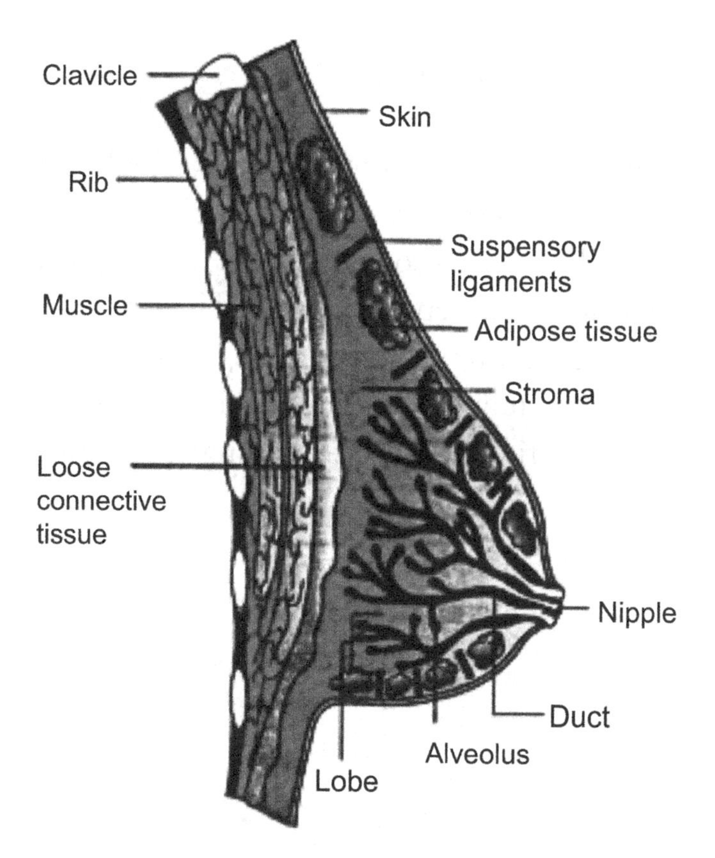

FIGURE 4: Mammary Gland Structure. Lactation ducts in each of the 15–20 lobes of the breast converge at the nipple. Each duct contains multiple alveoli that produce mother's milk. Reproduced from Hole's Human Anatomy & Physiology, 6th ed. WM C. Brown Publishers. Permission Pending.

NORMAL DUCTAL EPITHELIUM

Breast cancer begins with the transformation of a single epithelial cell lining a lactation duct. A cross section taken through a normal lactation duct is shown in Figure 5. The duct is completely open (*patent*) and is lined with a single layer of flat (*squamous*) epithelial cells. The single cell epithelium is seen to better advantage at high magnification in Figure 6. The ductal epithelium serves to completely separate duct contents from cells, structural matrix, and fluids circulating in the surrounding connective tissue.

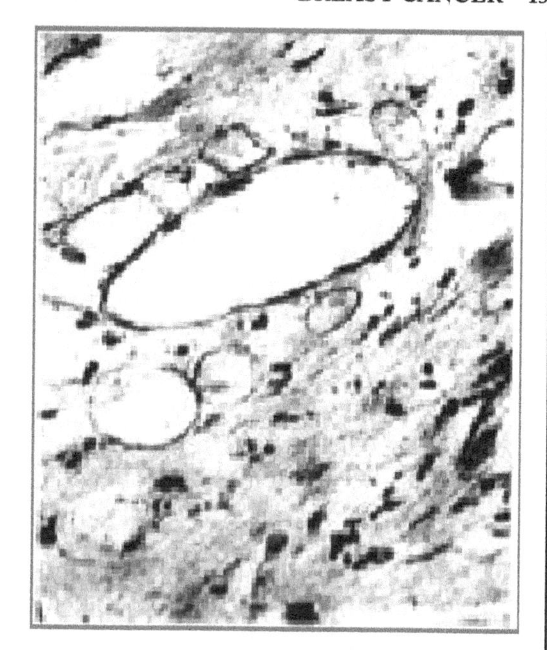

FIGURE 5: Normal Lactation Duct. A cross section through an empty breast lactation duct illustrates the extremely thin layer of squamous epithelial cells (dark line) profiling the duct. The duct is embedded in surrounding connective tissue containing cells, vessels, nerves, and an extracellular matrix of structural proteins. Courtesy of J.D. Pardee.

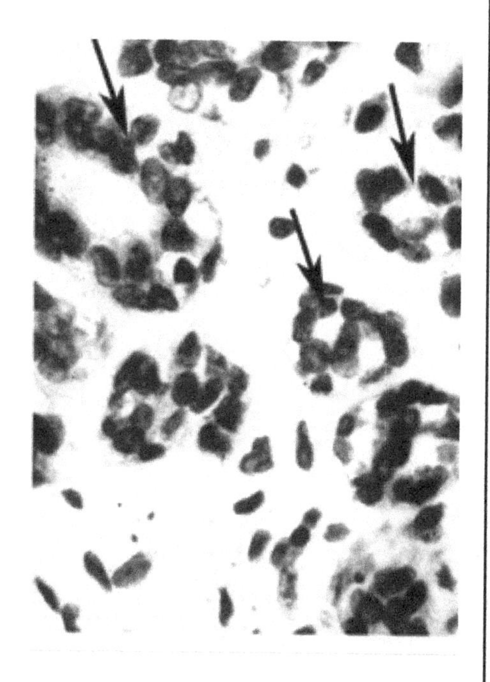

FIGURE 6: Lactation Duct Squamous Epithelium. High-magnification view of small lactation ducts cut in cross section. Arrows point to nuclei (blue) of squamous epithelial cells lining each duct. Surrounding connective tissue contains a few scattered cells (small blue nuclei) indigenous to the connective tissue. Courtesy of J.D. Pardee.

TRANSFORMATION: STAGE I

Now look at the lactation duct at an early stage of cancer (Figure 7). The single cell layer epithelium has been completely replaced by multiplying transformed cells that are filling the lactation duct. However, proliferating cells are still contained entirely within the duct and have not escaped into the surrounding connective tissue. This characterizes Stage I (transformation) of mammary carcinoma. The forming tumor is still benign because it is confined to the tissue of origin, in this case, the lactation duct epithelium.

A striking example of a benign tumor trapped inside the lactation duct and unable to progress to metastasis is seen in Figure 8.

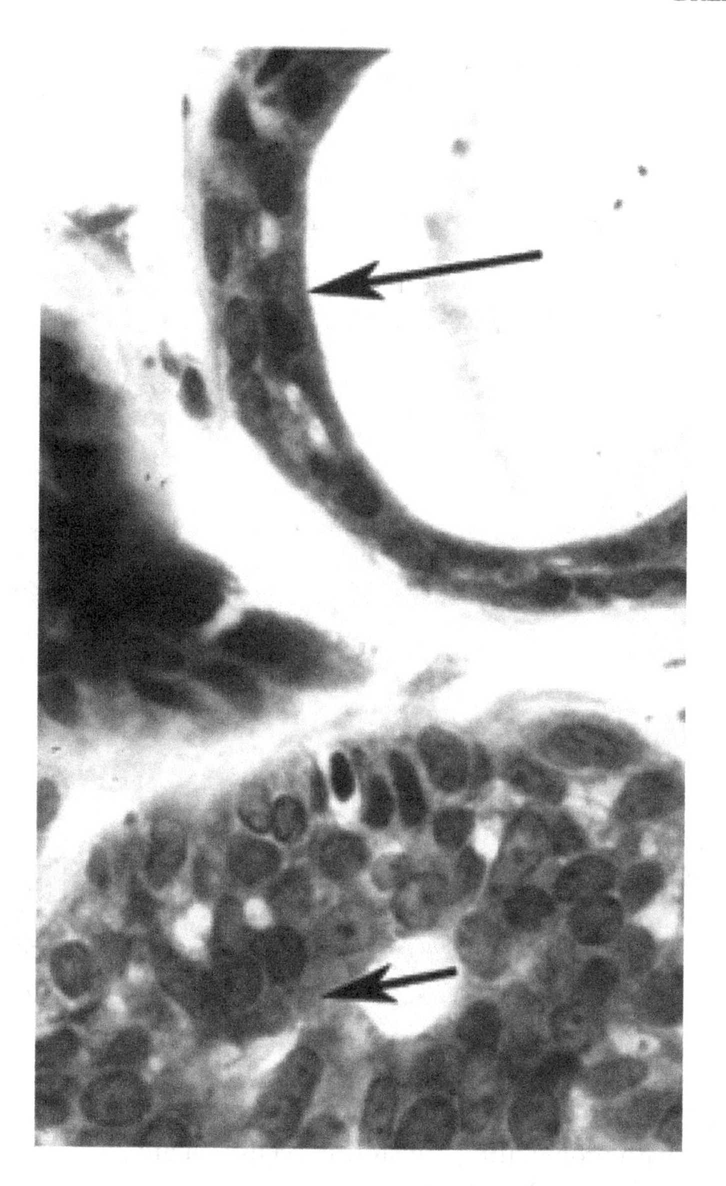

FIGURE 7: Lactation Duct Fibroadenoma. A fibroadenoma is a benign tumor contained within the lactation duct. The cancer cells are stained brown with a molecular marker that is specific for transformed cells (arrows). The histological section cuts through a single duct that appears multiple because the convoluted duct runs in and out of the section plane. The duct section showing transformed cells nearly filling the duct is closer to the initial site of transformation. Courtesy of J.D. Pardee.

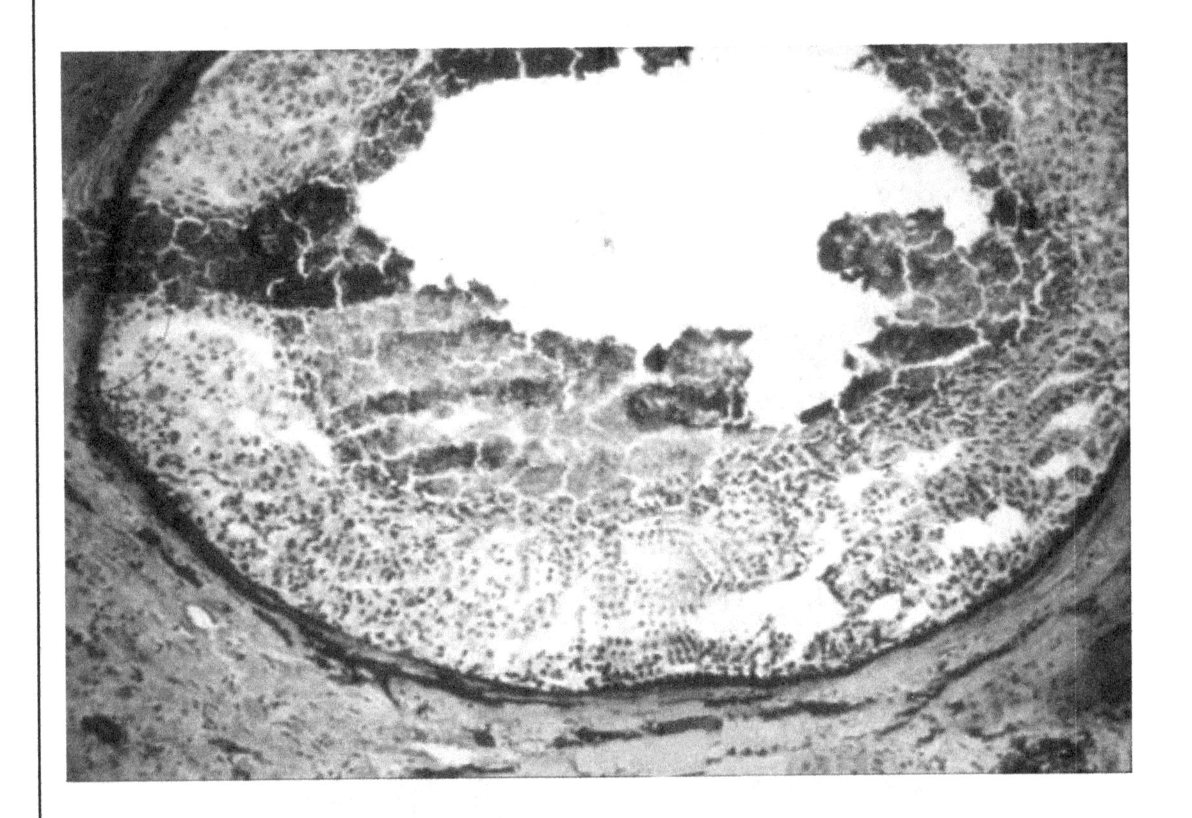

FIGURE 8: Ductal Carcinoma in Situ. This is a benign cancer of the breast that fills the lactation duct with proliferating transformed cells that are unable to invade the surrounding tissue. Courtesy of J.D. Pardee.

INVASION: STAGE II

The escape of proliferating cells from the lactation duct into the surrounding connective tissue marks the onset of invasion (Figure 9). Tumor cells have now acquired the ability to migrate, which requires genetic mutations that turn on proteins and enzymes that allow the cell to literally punch holes in the duct wall and crawl through into the connective tissue. Invasive cancer cells do not passively proliferate out of the duct but rather crawl like an amoeba into the connective tissue (*stroma*) of the breast.

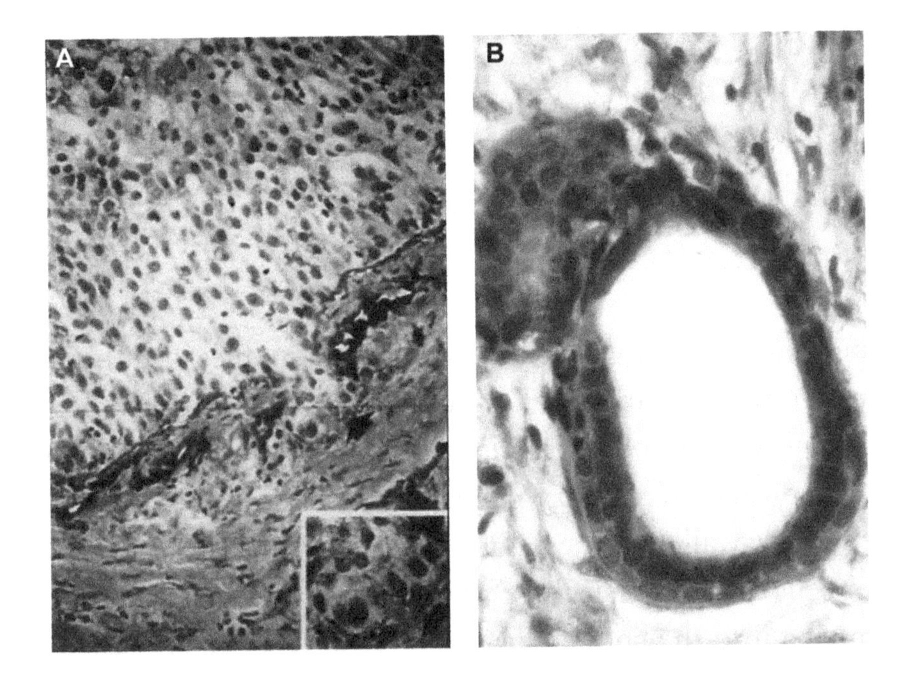

FIGURE 9: Ductal Cell Invasion. An example of invasion of a *ductal fibroadenoma* is shown. In this case, large, motile invasive cells (a) have penetrated the duct wall and are migrating into the adjacent connective tissue, where they continue to proliferate and migrate (b). Courtesy of J.D. Pardee.

MICROMETASTASIS

For late-stage cancers, it is estimated that the 15% and 34% mortality rates associated with Stage I and II diagnosis, respectively, are likely to indicate the presence of micrometastases (Figure 10) in connective tissue near the primary tumor site, in lymph nodes located under the arm (*axillary nodes*) and in distant organ sites. The 34% mortality rate accompanying Stage II disease diagnosis has been attributed to inadequate detection of occult metastases in axillary nodes and to incomplete removal of undetected micrometastases around the primary tumor during surgery.

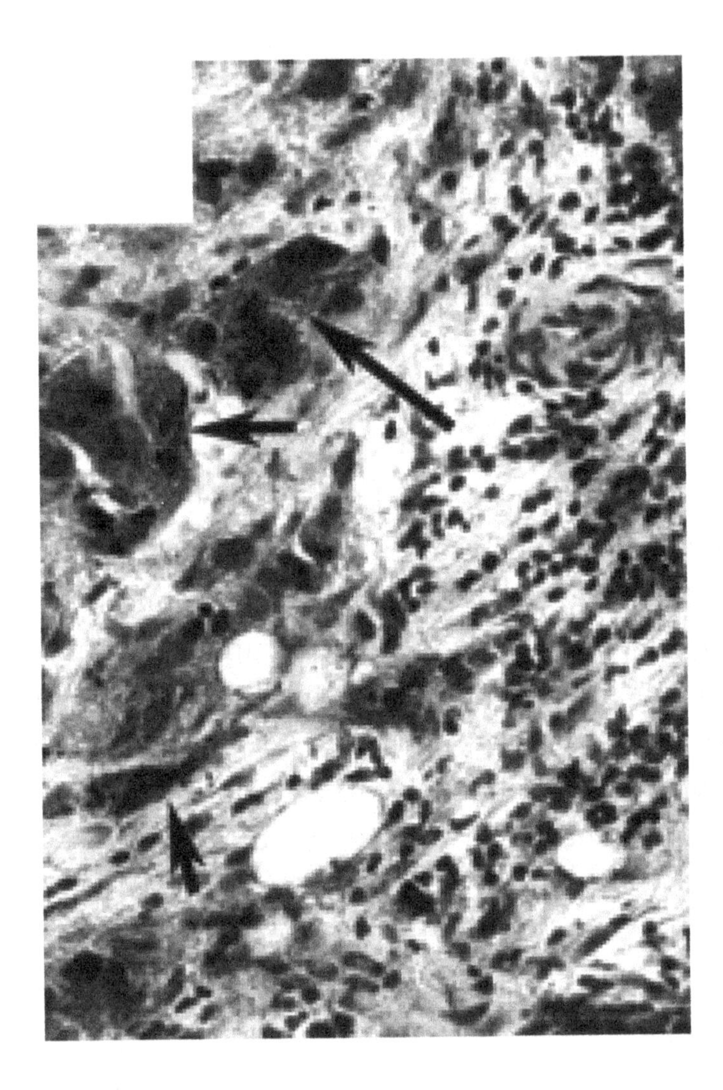

FIGURE 10: Micrometastases. Metastatic cancer cells (arrows) are stained brown with a molecular marker specific for transformed cell types. Motile, metastatic cells are quite large compared with the indigenous cells (small blue nuclei) of the connective tissue. This section was taken at some distance from the origin of the lactation duct tumor, confirming the importance of surveying tissue margins surrounding tumor surgical excisions for individual metastatic cells. Courtesy of J.D. Pardee.

Breast Cancer and Estrogen

A direct relationship between female hormones and breast cancer was established over 100 years ago, when Dr. George Beatson discovered in 1896 that removal of the ovaries (*ovarectomy*) led to remission of metastatic mammary carcinoma in about one third of his patients. The discovery of estrogen in 1923 and its identification in the circulation of premenopausal women cemented the relationship between estrogen and breast cancer. Estrus women between the ages of 24 and 55 years are at highest risk, roughly corresponding to the onset of ovulation (*menarche*) and cessation of ovulation (*menopause*), with the incidence of disease rising sharply with age up to the time of menopause. Contributing factors of early menarche (10–12 years old), late menopause (over 55 years), late pregnancy, obesity, and alcoholism are added risk factors. The risk of developing breast cancer following menopause drops sharply to roughly one sixth that of premenopausal women. These observations indicate that the transformation event in epithelial cells lining the lactation ducts is closely correlated with hormonal fluctuations accompanying ovulation. How can this be explained?

POINT OF ATTACK

Each menstrual cycle is a preparation for pregnancy. Ovaries develop an egg, the uterus prepares for embryo implantation, and the breast prepares for milk production. Fluctuations in the secretion of several hormones from the ovaries, principally estrogen and progesterone, trigger the cellular events creating tissue changes. In the breast, new milk-producing alveoli and supporting lactation ducts are built. New ducts sprout from established ones and lengthen, eventually forming alveoli at their ends. The act of ductal sprouting requires the conversion of nondividing epithelial cells in the parent duct into a rapidly dividing, proliferating epithelial cell. It is this dividing ductal epithelial cell that is the point of attack for transformation (Figure 11).

With each menstrual cycle, new lactation ducts may form. To support the metabolic needs of the new cells, extended capillary networks proliferate as well. Blood and fluids engorge the breast, causing swelling and pressure-sensitive pain. If pregnancy is not achieved, newly formed ducts quickly regress, in preparation for growth in the next cycle. Since the genetic changes causing transformation to continual cell division are most likely to occur in a dividing cell, the probability of spontaneous conversion to a proliferative state is highest in the dividing epithelial cells forming the nascent lactation duct. It is the high circulating levels of estrogen that trigger cell division in sprouting ducts. The window of opportunity for breast cancer opens with each estrus cycle.

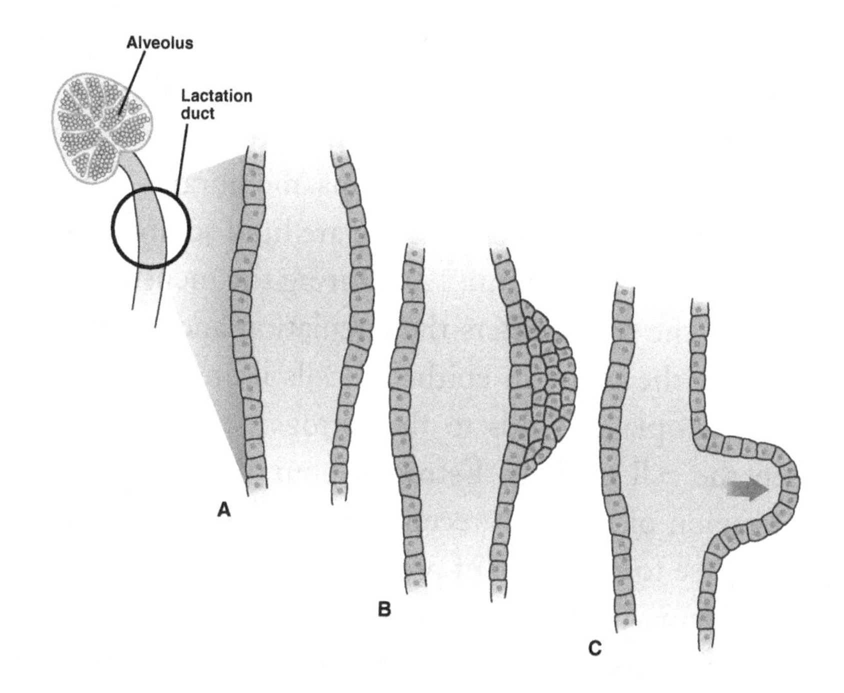

FIGURE 11: Lactation Duct Sprouting. In this model of lactation duct sprouting, epithelial cells containing an estrogen receptor (approximately 25% of ductal epithelial cells) are stimulated by estrogen from the ovaries to divide. As the stimulated cell continues to divide, a new lactation duct branch begins to bud from the existing duct. Courtesy of J.D. Pardee.

THE ESTROGEN RECEPTOR

How does estrogen trigger the division of target cells? The answer to that question began to be unraveled with the discovery of the estrogen receptor in 1966. The first thing to notice is that the estrogen receptor is located within cells, not on the cell surface plasma membrane. This is because steroids such as estrogen and progesterone are lipid soluble and pass freely through the cell membrane (Figure 12). During the menstrual cycle, estrogen secreted from the ovary enters the circulation and is carried to target tissues throughout the body. In epithelial cells lining the lactation ducts, estrogen in the cytoplasm binds to the estrogen receptor, which is then transported to the cell nucleus. Estrogen-bound estrogen receptor finds the promoter region of estrogen receptor target genes and activates transcription, giving rise to a number of proteins that turn on cell proliferation. Proliferation inducing proteins can act on the producing cell itself or can be secreted from the producing cell and act on plasma membrane receptors on neighboring cells (*paracrine activation*), including the producing cell itself (*autocrine activation*), as well as other cell types such as fibroblasts and vascular endothelium. The net effect is to stimulate sprouting of lactation ducts, restructuring of the supporting connective tissue, and activation of capillary growth (*angiogenesis*) in the neighborhood surrounding the sprouting duct.

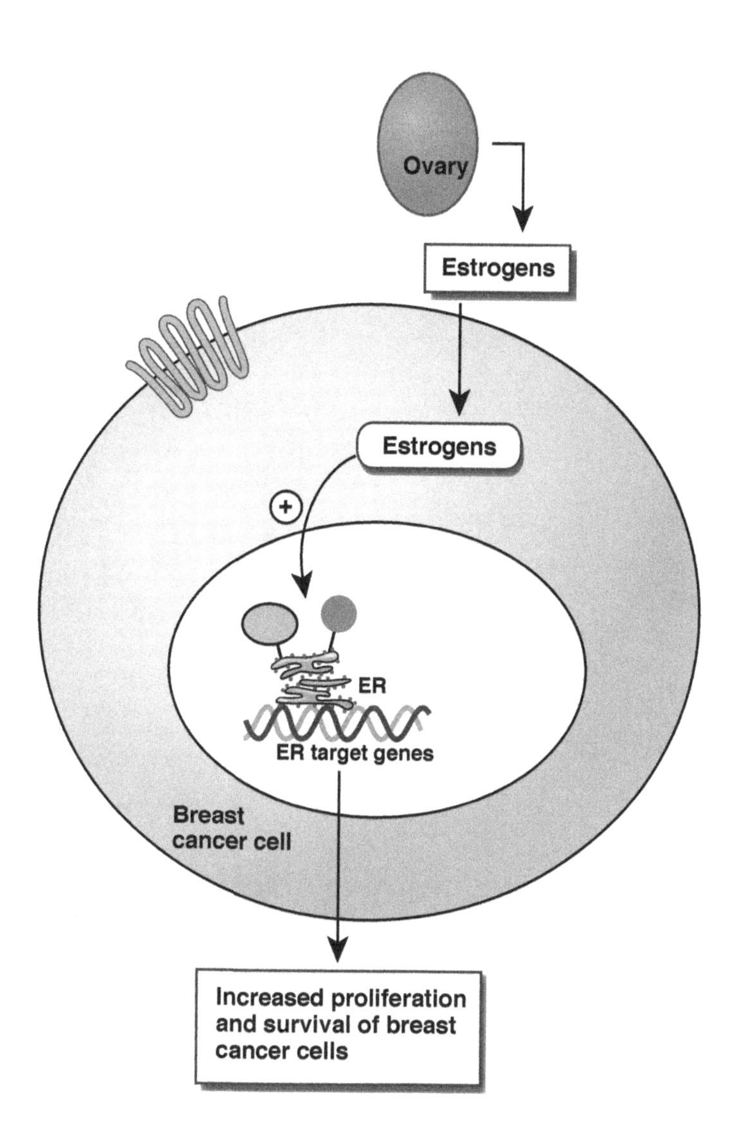

FIGURE 12: Breast Estrogen Receptor. A simplified model of estrogen receptor stimulation of cell proliferation is shown. In premenopausal women estrogen secreted from the ovaries enters breast lactation duct epithelial cells and binds to the estrogen receptor. Estrogen-bound estrogen receptor enters the nucleus, binds to the promoter region of estrogen receptor target genes, and activates transcription of protein factors that induce proliferation of the epithelial cell. Courtesy of J.D. Pardee.

Treating Breast Cancer

Approximately 80% of breast carcinomas are estrogen receptor-positive, i.e., they contain the estrogen receptor. Consequently, treatments of these cancers are focused on inhibiting estrogen-stimulated cancer growth and development. Let us now look at various treatments for estrogen receptor-positive breast cancer (Figure 13).

PREMENOPAUSAL ESTROGEN RECEPTOR POSITIVE BREAST CANCER: PRIMARY TUMOR WITHOUT METASTATIC DISEASE

We have been speaking of inducing proliferation of lactation duct epithelial cells, a perfectly normal process that requires estrogen. If transformation occurs, a primary tumor is established (tumor just means a growth with no implication of malignancy) and estrogen now takes on the role of stimulating and supporting cell proliferation within the tumor. Consequently, in premenopausal women, the first line of defense against primary tumor proliferation is to (1) surgically remove the primary tumor if possible, (2) prevent secretion of estrogen from the ovaries (*ovarian ablation*), and (3) pharmacologically block the estrogen receptor in the breast to prevent further proliferations.

1. Breast surgery. Once a tumor mass has been detected by breast exam (palpation) and x-ray mammography, a biopsy of the lump is taken. Thin sections of the biopsy material are stained to visualize cell morphology, and molecular markers are applied to determine the stage of transformation. Molecular markers are usually antibodies that bind specifically to proteins that are known to be present in the transformed cells but not present in normal cells. Examples of biopsy sections are presented in Figures 6–11. The stage of tumor development determines the therapeutic approach to be taken. If the tumor is well localized and benign (early Stage I, localized proliferation), surgical removal of the mass by *lumpectomy* is performed without further treatment. If advanced Stage I (extensive proliferation without metastasis) is detected, treatment is more aggressive. In addition to removal of the primary tumor, further treatments may include prevention of estrogen secretion from the ovaries and pharmacologic inhibition of the estrogen receptor.

2. Ovarian ablation. The goal of ovarian ablation is to prevent estrogen secretion. Temporary cessation of estrogen production in the ovary

can be accomplished by using *luteinizing hormone-releasing hormone antagonists*. Permanent loss of ovarian function requires removal of the ovaries.

3. Estrogen receptor inhibition. To block estrogen function in proliferating breast epithelium, drugs targeted to bind and inhibit the estrogen receptor are used. *Tamoxifen* is often the drug of choice, although the estrogen receptor antagonist *faslodex* is also used for women who have failed tamoxifen therapy.

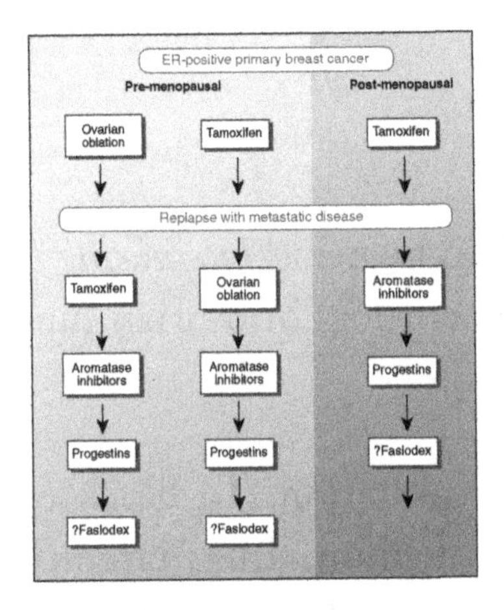

FIGURE 13: Treatment Regimes for ER Positive Breast Cancer. Sequential use of ovarian ablation (e.g. using luteinizing-hormone-releasing hormone antagonists), anti-estrogens (typically tamoxifen, although the 'pure' estrogen receptor (ER) antagonist faslodex is also used for women who have failed tamoxifen therapy), progestins and/or aromatase inhibitors are now commonly used in the management of hormone-sensitive breast cancer. Courtesy of J.D. Pardee.

PREMENOPAUSAL BREAST CANCER: RELAPSE WITH METASTATIC DISEASE

When patients initially present with Stage II invasive carcinoma or have relapsed into metastatic disease after initial treatment for Stage I proliferation, a complex and aggressive treatment regime is followed (Figure 13).

1. *Surgery* includes removal of the affected breast (*unilateral mastectomy*) or both breasts (*bilateral mastectomy*), depending on the extent of metastasis. In addition, partial or complete removal of the axillary lymph nodes is performed if cancer cells are detected in the nodes. While breast reconstruction is now a routine option following mastectomy, axillary node removal is fraught with potentially negative side effects as well as controversy over its therapeutic benefit. The principal problem with removal of axillary nodes is the resulting *lymphedema* of the arm. Lymph flows in the arm from the fingertips to the axillary nodes in the armpit to the neck, where it enters the venous blood circulation. Surgical removal of the axillary nodes interrupts normal lymph circulation, causing lymph to collect in the arm, resulting in extensive swelling and pain from the buildup of fluid pressure. Unlike breast reconstruction, there is no surgical reconstruction capable of restoring normal lymph flow through the *axilla* (armpit). Depending on the extent of node removal and skill of the surgeon, patients risk enduring chronic and permanent pain in the

affected arm, sometimes with tragic consequences. There is also no little controversy over the effectiveness of node removal in increased survival. Detection of breast cancer cells in the axillary nodes indicates the onset of Stage III cancer, since metastasis has progressed to the blood circulation and lymph ducts draining the breast. Since nodes function to filter lymph free of cancerous cells and to mount an immune response against transformed cell types, points of view differ with respect to the efficacy of node removal.

2. In metastatic disease, removal of circulating estrogen by ovarian ablation must be complemented by breast estrogen receptor inhibition with tamoxifen and faslodex. Furthermore, the discovery of another source of intracellular estrogen derived from circulating steroids has precipitated the development of drugs that inhibit the enzyme *aromatase* (Figure 14). Aromatase is found in the cytoplasm of breast epithelial cells. It converts steroids to estrogen, thus supplying additional estrogen to drive the proliferation and survival of transformed lactation duct epithelium. When the primary source of estrogen from the ovaries is eliminated, aromatase continues to supply it. The principal source of steroids is the adrenal gland, situated atop the kidney. Circulating steroid hormones including androgens such as testosterone enter the ductal epithelium, bind aromatase, and are converted to estrogen. In both premenopausal and postmenopausal women, low levels of estrogen are maintained in breast cancer cells by aromatase, necessitating administration of aromatase inhibitors.

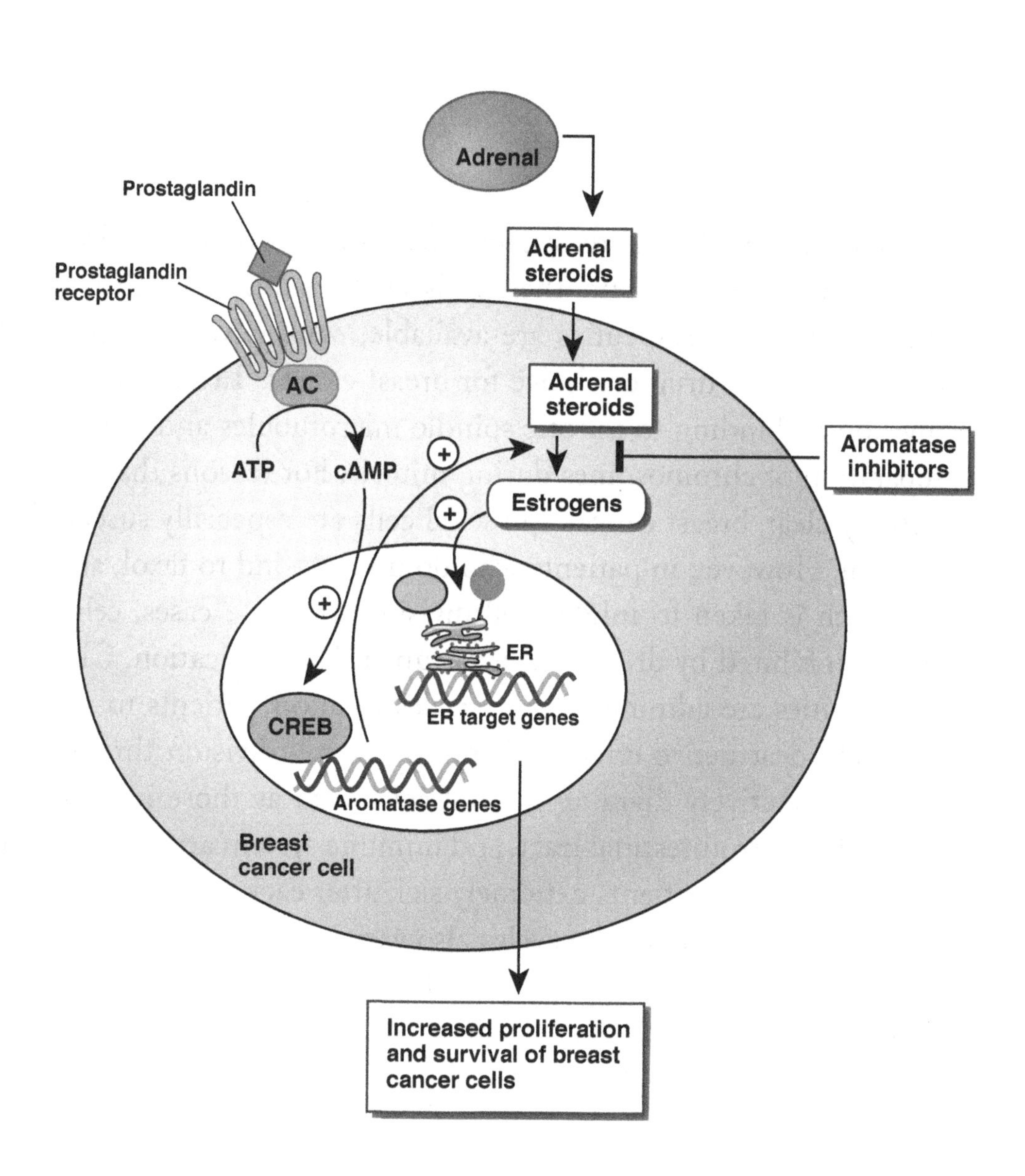

FIGURE 14: The Aromatase Pathway. Courtesy of J.D. Pardee.

3. Chemotherapy. When breast cancer has progressed to Stage IV (disseminated disease), chemotherapy is employed. Although a spectrum of chemotherapeutics are available, *taxol* is widely acknowledged to be the drug of choice for breast cancer. Taxol inhibits cell division by binding to mitotic spindle microtubules and preventing separation of chromosomes during mitosis. For reasons that are not entirely clear, breast cancer epithelial cells are especially susceptible to taxol. However, in patients who do not respond to taxol, another approach is taken to inhibit cell division. In these cases, cell division is inhibited by drugs that interrupt DNA replication. Chemotherapeutics are administered in cycles to allow patients to recover from the destructive effects of preventing cell division throughout the body. Actively dividing normal cells such as those in hair follicles, the gastrointestinal tract, and immune system are also severely affected, leaving patients extremely sick after each round of chemotherapy. The timing of the cycles also varies with the type of cancer being treated, i.e., fast growing or slow growing, the disease stage, and the general health of the patient.

Conclusion

This short introduction to breast cancer is not intended to be encyclopedic or even thorough. Rather, the aim is to render a conceptual molecular vision of how the disease begins, progresses, and ends. Only by seeing in our mind's eye the "molecular movie" reeling out the story of cancer do we begin to understand it. It is by intuitive understanding and not mere factual knowledge that we dance from around the ring to the secret center.

References

FURTHER READING

Alberts, B., Johnson, A., Lewis, J., Raff, M., Roberts, K., and Walter, P. (eds.) (2002). Molecular Biology of the Cell, 4th ed., New York: Garland Science.

Berg, J., Tymoczko, J., and Stryer, L. (2007). Biochemistry, 5th ed., New York: W.H. Freeman and Company.

Cairns, J. (1978). Cancer: Science and Society, San Francisco: W.H. Freeman.

De Vita, V.T., Hellman, S., and Rosenberg, S.A. (eds.) (2000). Cancer: Principles and Practice of Oncology, 6th ed., Philadelphia: Lippincott, Williams and Wilkins.

Greaves, M. (2000). Cancer: The Evolutionary Legacy. Oxford: Oxford University Press.

Lathem, E. (ed.) (1969). The Poetry of Robert Frost, New York: Holt, Rinehart and Winston.

Pollard, T., and Earnshaw, W. (2007). Cell Biology, 2nd ed., Philadelphia: Saunders Elsevier.

RESEARCH ARTICLES

Baselga, J., and Albanell, J. (2001). Mechanism of anti-HER2 monoclonal antibodies. *Ann. Oncol.*, 12, 535–541.

Cairns, J. (1985). The treatment of diseases and the war against cancer. *Sci. Amer.*, 253(5), 51–59.

Chambers, A.F., Naumov, G.N., Vantyghem, S., and Tuck, A.B. (2000). Molecular biology of breast cancer metastasis; clinical implications of experimental studies on metastatic inefficiency. *Breast Cancer Res.*, 2, 400–407.

Doll, R., and Peto, R. (1981). The causes of cancer: quantitative estimates of avoidable risks of cancer in the United States today. *J. Natl. Cancer Inst.*, 66, 1191–1308.

Fialkow, P.J. (1976). Clonal origin of human tumors. *Biochim. Biophys. Acta*, 458, 283–321.

Folger, P.A., Berg, W.J., De Jesus, Z., Fong, Y., and Pardee, J. (1999). A mammalian homolog of *Dictyostelium* severin, m-severin, replaces gelsolin in transformed epithelial cells. *Cancer Res.*, 59, 5349–5355.

Folkman, J. (1996). Fighting cancer by attacking its blood supply. *Sci. Amer.*, 275(3), 150–154.

Hanahan, D., and Weinberg, R.A. (2000). The hallmarks of cancer. *Cell*, 100, 57–70.

Klein, G. (1998). Foulds' dangerous idea revisited: the multistep development of tumors 40 years later. *Adv. Cancer Res.*, 72, 1–23.

Kreitman, R.J. (1999). Immunotoxins in cancer therapy. *Curr. Opin. Immunol.*, 11, 570–576.

Malpas, J. (1997). Chemotherapy. In Introduction to the Cellular and Molecular Biology of Cancer (L.M. Franks, N.M. Teich. eds.), 3rd ed., pp. 343–352. Oxford: Oxford University Press.

Nowell, P.C. (1976). The clonal evolution of tumor cell populations. *Science*, 194, 23–28.

Peto, J. (2001). Cancer epidemiology in the last century and the next decade. *Nature*, 411, 390–395.

Ridley, A. (2000). Molecular switches in metastasis. *Nature*, 406, 466–467.

Ruoslahti, E. (1996). How cancer spreads. *Sci. Am.*, 275(3), 72–77.

Soriano, Z., and Pardee, J. (2004). M-34 actin regulatory protein is a sensitive diagnostic marker for early and late-stage mammary carcinomas. *Clin. Cancer Res.*, 10, 4437–4443.